Fa science

All About Fart: Farting, Passing Gas And Flatulence

By

Nishi Singh

Disclaimer:

Whilst every care is taken to ensure that the information in
this book is as up-to-date and accurate as possible, no
responsibility can be taken by the author for any errors or
omissions contained herein. Some images used in this book
are in the public domain compiled from various image
repositories. Research for the book has been done using
reliable sources and the author's own person experience. In
the event where any fact or material is incorrect or used
without proper permission, please contact us so that the
oversight can be corrected. Responsibility for any loss,
damage, accident or distress resulting from adherence to
any advice, suggestions or recommendations is not taken.

ISBN-13: 978-1519210593

ISBN-10: 1519210590

Other books by the Author:
http://www.amazon.com/author/nishisingh

Table of contents

1. Introduction

Welcome to this book about flatulence. For those of you who aren't familiar with this word, it means to build up gas in the digestive tract and expel it through the anus. You probably know the simpler term known as "fart." When we fart we are releasing gas from our anus that got created in the digestive tract after it broke down the food we ate. Since everyone eats, then everyone farts. There is no way around it. Don't believe the myth that only guys fart and that girls never fart because it is not true. All human beings fart, whether you are a guy or girl. So when you are in school and you see a really pretty girl, don't assume she doesn't fart because she is pretty. Trust me, she farts! If you got to spend a lot of time around her then you might notice her going to the bathroom frequently or letting out a little squeaky fart while thinking nobody will notice. Haven't we all tried to mask our own farts by clearing our throat or making some sound to distract people from it? The answer is an absolute YES.

It is kind of funny to think that everybody farts. We don't think about this too much because farting is often seen as embarrassing by our society. That is why we try so hard to hold in our farts when we are around other people. But all this is going to do is build up even more gas inside of your stomach, which will have to come out sooner or later. Then you are bound to release a monstrous fart that will be even louder and even smellier than it would have been before. So you might as well fart when your body tells you that you have to fart. However, this doesn't mean that you purposely try to make the loudest and smelliest fart possible. Just let your farts come out gradually without forcing them out. As for the smell, there isn't a whole lot you can do about that unless you purposely eat foods that

make stinky farts. These types of foods will be
discussed in this book in case you are really curious.
The idea is to refrain from eating these foods and
substitute them with foods that produce less smelly
farts. That way you don't have to make everyone
sick around you the next time you have to let one
go.

This book is meant to be both fun and educational
at the same time. It will explore in depth about why
flatulence exists in the first place and what exactly
happens in our body to cause gas to get created.
You will also learn about why farts create different

sounds and smells. There are very good reasons for why farts smell and sound the way they do, so it is important to learn these reasons. Not only will this help you manage your own farts better, but you can help educate others on their own farts as well. It may seem like an uncomfortable topic to talk about with your friends and family, but it doesn't have to be. Flatulence is a biological function that could even be caused by diseases or cancers in the body. Although these are rare causes of flatulence, it is important that you understand it completely so you can judge for yourself if your excessive fart problem is something serious. As for those who don't have an excessive fart problem, you will one day. We all have fart problems at one point or another. But hopefully after you read through this book you can appreciate your farts for what they are without always thinking badly about them.

2. What is fart and flatulence

Fart and flatulence (flach-u-lens) are two words that are used synonymously. Farting basically describes when we actually let out gas through our anus after it has already built up inside of our body. Flatulence describes the overall process from the beginning of the build up to the end. The scientific definition of flatulence is having excessive amounts of intestinal or stomach gas that gets produced during digestion and then released through the anus while giving off an odor and sound. To really understand where the word comes from, you have to study the root word "flatus." This is actually a Latin word that means "breaking wind." Medical doctors use this word to describe when someone generates gas in their bowels or stomach. All of the excess gas build up occurs in the digestive tract, which starts from your mouth and ends down to your anus. Everything in between these two areas of the body is what helps pass food along, so that it can be broken down and converted into energy. The excess waste that gets produced as a result will produce waste gas in the intestines, which is where farts are born.

The four main causes of having excess gas are swallowed air, being lactose intolerant, imperfect absorption of foods, and the breakdown of undigested foods. Sometimes these can even cause flatulence through the mouth instead of the anus, which is known as burping. People often don't think of burping as being in the same category as farting, but the gas that produces both symptoms actually come from the same place (the digestive tract). So when you are burping you are really farting through your mouth. However, the smell and sound of bumping varies a bit from the smell and sound of farting. The gas farts that come out your anus contain a little bit of sulfur, which make them smell worse. But you will learn more about that as you read through this book.

The important thing to know about flatulence is that it comes from a healthy, but complex ecosystem of bacteria that resides inside of our intestines. That right's we actually have living organisms inside of our intestines which feed on the food that passes through it. These organisms are bacteria that survive off the carbohydrates in our food. When the bacteria break down the carbohydrates for the body, this produces the gas that eventually becomes a fart. These same bacteria are responsible for helping the body to absorb fatty acids and vitamins from the foods, which supports the health of our immune system and maintains the lining of our colon. So the next time you end up farting nonstop, this should serve as an indicator to you that your body has successfully broken down the foods you ate in order to preserve your overall health. This is what makes flatulence a healthy and natural function that should be admired instead of looked down upon.

Here are some interesting facts about flatulence that will probably surprise you. On a scientific level, people produce up to 1,500 milliliters of gas every day in their digestive tract. If you want to imagine this just think about a 2 liter soda bottle that you see in the store. 1,500 milliliters would equal ¾ of the entire 2 liter soda bottle. You can expel this much gas from the body in about 20 farts. This averages out to 75 milliliters of gas per fart. These measurements may freak you out because when you imagine all of that gas filling up ¾ of a 2 liter soda bottle then it may seem excessive. But you have to remember that you fart constantly throughout the day and that each time you expel a fart, it contains a lot of gas in it. You just don't really think about it because you can't see the farts and often times you can't smell them either.

3. The digestive system and flatulence

When we talk about "gas" we aren't talking about the liquid fuel that you fill up your car with. This gas is an air-like fluid that cannot be seen with the naked eye. Flatulence is nothing more than intestinal gas that builds up in the digestive tract. The digestive tract is a big muscular tube that starts from your mouth and goes all the way down to your anus, which is that hole in the middle of your butt. The digestive tract is responsible for breaking down the food you eat so that your body can extract its nutrients. In order to do this, the digestive tract releases enzymes and hormones to make this possible. This creates a nice mixture of gas chemicals such as nitrogen, oxygen, carbon dioxide, hydrogen and in some cases even methane. Gas from the digestive tract can actually leave your body in two places; the mouth and the anus. The gas that leaves your mouth may smell, but it doesn't smell nearly as bad as the gas that comes out of your anus. The reason for this is because this gas also has small amounts of sulfur in it, which you may know to be a chemical that contains a very bad odor. This odor is the predominant reason for why farts smell so bad. Since burping doesn't contain this sulfur, the gas that comes out of your mouth is not as foul smelling as the anus farts.

So what causes gas in the first place? Believe it or not, gas forms in the digestive tract from two things; swallowing air and food breakdown by bacteria in the large intestine. People often forget about how swallowing air makes you fart. They just think of the nasty foods that cause gas and nothing else. But the truth is that we all swallow small amounts of air every time we drink and eat. The amount of air that gets swallowed depends on a variety of things, like how fast someone eats and drinks. Let's say you are really hungry and you order a big cheese pizza from your local pizza joint. If you eat that cheese pizza fast, not only will you be eating a lot of food but you will be swallowing a lot of air as well. Basically the more you eat the more air you consume, but you consume more air if you eat fast. Other ways to swallow air include smoking, chewing gum, sucking on candy, drinking carbonated drinks and wearing loose fit dentures. All of these things can contribute to gas accumulating in the digestive tract.

The average person farts about 20 times per day. People with flatulence usually have excessive amounts of gas in their digestive tract, which makes them fart more than 20 times per day. This is caused by gas build up in their large intestine and stomach, which causes them to bloat. In rare situations, bloating from excessive gas could be caused by diseases in the intestines such as rapid gastric emptying or intestinal obstruction. But most of the time it is just due to people consuming too many carbohydrates at once. This puts a damper on the digestive tract and causes a gas overload in your intestine and stomach. The best way to avoid this from happening is to eat foods with fewer carbohydrates. You should also eat your foods a lot slower as well. If you are hungry then just try to eat at a normal pace without stuffing yourself silly. Remember that your digestive tract can only process so much food at once. If you think about the food you eat being inside your digestive system then you will probably understand all the work it has to do to break down the food. This will make you want to eat in moderation, which is what people are naturally meant to do.

4. Why do we fart - causes of flatulence

This book has already touched upon the causes, but now it is time to examine them more closely. You already know that flatulence is when excessive gas builds up in the digestive tract, which makes up the stomach, esophagus, small intestine, large intestine and colon of the body. The two main reasons for gas buildup could be either from swallowing air while eating or from the breakdown of undigested foods by bacteria inside your colon. The gas from swallowed air moves into your small intestine. People tend to expel this gas by burping out the air through their mouth. This gas mainly contains oxygen, carbon dioxide and nitrogen. The other kind of gas that gets created in the digestive tract ends up in your large intestine and eventually gets expelled through the rectum. This is the gas that contains hydrogen and methane compounds, which have elements of sulfur in them. Sulfur is what makes gas smell like rotten eggs.

Let's look at swallowing air first. The scientific name for swallowing air is "aerophagia," which could refer to accidently or deliberately swallowing air. In the case of flatulence, the term refers to people who swallow air while they are eating or just swallowing involuntarily out of habit. When people are not eating, they tend to swallow air from doing certain activities like chewing gum, using tobacco, sucking on hard candy, drinking too fast, drinking carbonated beverages, hyperventilation or loose dentures. The unfortunate part is that nobody really notices when they are swallowing in air while conducting these activities. Either that or they just don't think about it because their mind is preoccupied with other thoughts. Since breathing and swallowing tend to be involuntary actions, we never really know how much air we end up swallowing in a given day. The good news is that belching up your swallowed air won't smell as bad as farting because there are no sulfur chemicals in the air you swallow.

People fart numerous times throughout the day. Sometimes these farts come from the nasty foods we eat. Perhaps you ate spicy tacos or baked beans. How about a double patty cheese burger? Gas forms after eating all kinds of foods, but some foods cause it more than others. Basically, the foods that are harder for the body to break down will cause flatulence. For example, if you eat foods that contain lots of fiber and complex carbohydrates then you will likely experience a lot of excess gas. That is why people always joke about Mexican food making you fart a lot more because Mexican food tends to have a lot more bean ingredients in them. Beans, such as kidney beans and lentils, contain high amounts of fiber and carbohydrates will make you fart up a storm if you consume a lot of them.

The body is not able to digest carbohydrates, like fiber, starches, and sugar, in the small intestines because there is a shortage of enzymes there. This causes the undigested food to pass into the large intestine from the small intestine. At this point, bacteria living inside your large intestine break down this food and produce carbon dioxide, hydrogen, and sometimes methane gases as a result. These are the smelly gases that come out your rectum and cause you to have a rotten egg stench around you. Now this doesn't mean you did anything wrong with your eating. In fact, fiber filled foods are important to maintain the functionality of the digestive tract so it can process food faster. But the downside is that you have to put up with smelly farts afterwards, which can be awkward if you hang around a lot of people.

5. Chemical composition of fart

There are trillions of chemical reactions taking place inside of our bodies that keep it alive, a process called "metabolism". It just so happens that farts are the funniest chemical reaction in our bodies. Not only that, they can be the most embarrassing! People always make a big deal over the sound and smell of farts. Usually the loudest and smelliest farts get the most attention. Often times the smelly farts might not even be loud, and vice versa. This might seem complicated to a layman person who has never studied the chemical composition of farts before. Why would they? Fart science is not exactly a topic that most people are interested in and it is certainly not taught at any schools in the country. But now you are in luck because you are going to learn the specific chemical composition of farts in this chapter.

A lot of our gas simply comes from ingesting air. This means the composition of our gas is merely air. The majority of these farts contain nitrogen in them, which is the main type of gas found in the air we breathe. You will also find a generous amount of carbon dioxide in the air as well. These two gases are what stimulate most farts from the air we breathe. Now let's look some general percentages of the gases found in air. There is 20% to 90% nitrogen, 0% to 50% hydrogen, 10% to 30% carbon dioxide, 0% to 10% oxygen and 0% to 10% methane. The methane and hydrogen are the only two gases that are flammable. If you have ever seen the movie "Dumb and Dumber" where Jim Carrey tries lighting his farts on fire, you now understand the science behind it. The theory is that the methane and hydrogen gases that make up the fart will flame out if you hold a lighter directly in front of your anus. While this may cause the flame to torch outward slightly, there generally isn't enough gas to cause any huge flames because the amounts of methane and hydrogen are so small. But just to be on the safe side, never try to light your farts on fire because you don't want to risk burning your anus. This will cause your farts to hurt.

Now let's look at another kind of fart that has more complex chemistry. These are farts that come from bacterial production or the digestion of food in our digestive tract. There are all sorts of nasty chemicals and gases that make up this function. For starters, meat digestion will produce indole and skatole chemicals that contribute to smelly farts. But the chemicals produced that create the rotten egg smell include methanethiol, dimethyl sulfide and hydrogen sulfide. These are all compounds of sulfur, which are both smelly and flammable at the same time. There is one story I read somewhere that this person eat a big juicy taco and then tried to light his farts on fire 30 minutes later, he actually saw some flames in the air. We now that this was due to all the sulfur in his farts. You will also get sulfur in your farts from eating complex carbohydrates because the bacteria in your intestines that feed on these carbohydrates will produce sulfur as a result.

If you want to control flatulence then you have to think about the chemistry behind the foods you eat. That is the only way to reduce the smells and possibly even the sounds of your farts. However, it will be impossible to always have odorless smelling farts because there will be a time when feces residue in your rectum will cause your farts to smell.

6. Why does fart smell so bad

There are all different kinds of fart smells, but one thing we can all agree on is that farts smell bad for the most part. People never produce pleasant smelling farts. If anyone thinks they are pleasant then they should have their head examined because something might be mentally wrong with them. But despite all farts smelling bad, there are certain farts that smell worse than others. It all depends on the foods that were eaten by the person producing the fart.

However, not all gas produced in the body causes a smell. In fact, 99% of all bodily gas is completely odorless. This 99% comes from the methane, carbon dioxide and hydrogen gases in the large intestines of our body. Have you ever noticed a time when you farted and didn't smell anything afterwards? This is likely due to your fart being caused by a combination of these gases without any sulfur added to the mix. Sulfur, or hydrogen sulfide, resides in the 1% of our farts that do stink. This is the main cause for the terrible rotten egg smell that has become synonymous with stinky farts. If you have ever bought sulfur-based acne medication or visited the mud volcanoes at Yellowstone National park then you will already be familiar with this smell. But how do sulfurous gases get produced inside the body? The way it develops is through the consumption of complex carbohydrates because many have sulfur compounds in them. Foods with complex carbohydrates include onions, Brussels sprouts, cauliflower, beans, dairy, and broccoli. Once the foods pass through the digestive system, bacteria inside the large intestine end up feeding off these complex carbohydrates. Then as the other chemicals in the body are helping to break down these foods, sulfur gets produced from the feasting bacteria and mixes with the other gases that form.

Eventually, these gases find their way out of your anus and produce stinky farts that make everyone nauseous around you.

There is no real way to get rid of these stinky farts because they are a natural biological function of the body. The best thing you can do is take preventive action in order to reduce the chances of having stinky farts. This means cutting out complex carbohydrates, like fiber, and eating cleaner meals that contain vegetables in them. However, you still need fiber in your diet in order to maintain the health of your digestive tract. Therefore, there is not much you can do to reduce stinky farts forever. You could take a break from eating these indigestible foods, but you will never be able to cut them out completely from your diet. If you do then it could actually cause you to have serious health issues, such as constipation, obesity, cancer, and cardiovascular disease. So when you weigh these effects of having no fiber versus the effect of having stinky farts with fiber, which one wins? Most logical people would say the fiber wins because the health of your body will be maintained.

On a final note about smelly farts, always make sure you clean out your rectum of any excess feces that may exist from the last time you went to take a poo. You may have a dirty rectum if you keep smelling rotten eggs when you fart, even though you have eaten the right foods to prevent the smell from occurring. To find out for sure, go into the shower and clean your rectum out with the shower water. Don't just rely on toilet paper to wipe away the feces residue because paper won't always do the trick. The only way the residue will get removed completely is if it is washed out with water. After this is done, you should notice your farts smelling much better.

7. What causes farting sound

There are all kinds of fart sounds. In fact, fart sounds are so popular that there are even apps you can download for your Android or iOS smartphones that can generate farting sounds by simply tapping buttons. But when it comes to a physical fart sound, this comes directly from the anus. It is the sound that occurs when gas passes through your rectum and out your anus. The reason sounds are made from farting is because the passing gas ends up vibrating the anal sphincter, which is a muscular ring that surrounds the anal canal. The more gas that tries to pass through the anal canal at once, the more the anal sphincter will vibrate and create loud obnoxious sounds. For example, if you just ate a big Mexican meal that consisted of tacos, chips and salsa, you will likely generate a lot of gas build up in your digestive tract from this food. This means excessive amounts of gas will try to pass through your anal canal at once. As a result, you will produce loud smelly farts about every 60 seconds.

In addition to the anal sphincter causing fart sounds, these sounds also get produced from closed buttocks. People who have tight butt cheeks that completely cover their anus will tend to have louder farts because the gas that comes out of their anus has to pass through the butt cheeks before they can hit the open air. You have probably seen people create this effect for fun with their hands and wrists. To make a fart sound manually, just put your two wrists together and extend your hands outward. Then put your wrists up to your mouth and blow in between them and then close it quickly. This is how jokesters create their fart sounds, which end up being pretty loud. But what is special about this effect is that it represents what happens when gas comes out the anus and tries to pass through two enclosed areas. Instead of it being your wrists, it ends up being your butt cheeks. So practice blowing into your clenched wrists and see what kind of fart sounds you can make. The harder you blow, the louder the sound. This would signify how a lot of gas passing through the anal sphincter at once will create a louder sound because of the constant vibrating that it goes through.

The speed of gas is just as important as the quantity of the gas passing through the anal sphincter.

Studies show that farts can travel up to 10 feet per second. So when it is coming out of your body, you can already imagine how fast the gas must be traveling when it goes through your anus. This is why it is able to create such a loud sound. If the gas were to just slowly puff out of your anus then you probably wouldn't hear anything at all. In fact, there are times went farts do this. These are called silent farts, which might sound like wind blowing in the air. But they won't sound like a bomb went off like loud farts normally do. However, silent farts can still contain the same rotten egg stench that loud farts have. Gas passing through the large intestine will usually have some trace of sulfur in it. So if the gas comes out slowly, then chances are the stench won't go very far and will cause you to have to smell your own farts. On the upside, at least your farts won't spread around the room like a cloud of rotten eggs and make everybody uncomfortable. That way you can maintain some dignity and not feel embarrassed.

8. Is smelling fart is good for you

Did you ever wonder if it is healthy to smell other people's farts or possibly even your own farts? You probably have not because this would be quite unusual. As human beings, we tend to react to something negatively when it makes us feel comfortable. When it comes to smelling farts, it is never a comfortable feeling even if they are your own farts. Some people may not mind the smell of their own farts, but they will certainly mind the smell of other people's farts. Either way, people don't usually think about farts any further than how bad they smell. What about their health effects on the body? A natural reaction to this question might be that smelling farts are bad to smell because it makes you uncomfortable when you smell them. But sometimes the things that make us uncomfortable are not necessarily bad for us. If you look at healthy foods, exercise, and maintaining relationships with people, these are all things that can make us uncomfortable. However, it doesn't mean we are going to quit them because we know the rewards outweigh the discomfort. Smelling farts are the same thing. They are actually beneficial to smell, even though they are difficult to tolerate.

According to a study published in the medical journal entitled "Medicinal Chemistry Communications," hydrogen sulfide gas can actually help treat people with certain diseases. They did go on to mention that breathing in large doses of sulfide gas can be dangerous to your health. But if you were to just take a small sniff of sulfide gas on occasion, it could actually reduce your risks of getting a stroke, heart attack, arthritis, dementia and even cancer. The way sulfur is able to do this is by preserving mitochondria in the cells of the body (cells are the basic unit all living things – every part of the body is made of cells). Normally when you have a disease, the mitochondrial cells end up getting damaged. This causes the energy and respiration production of the body to get reduced. But with the occasional sniff of sulfur filled farts, it can actually help repair the mitochondrial cells and keep them operational to preserve the health of the body.

Now before you start going out and sniffing people's butts, don't get carried away with this. Like previously mentioned, inhaling too much hydrogen sulfide is bad for your health. You need to sniff it in small amounts over an extended period of time. That is the best way you will get the health benefits that were mentioned. As far as treating cancer goes, don't get the wrong idea about this. Hydrogen sulfide is not the cure for cancer. If it were then scientists surely would have published their findings on it by now. But what you should understand is that hydrogen sulfide is great at reducing your chances of getting cancer along with the rest of the health problems that were mentioned. Therefore, you need to find a way to implement fart smelling into your daily routine. Chances are if you work at a fast food restaurant or some other restaurant where you are around customers all day, then you are probably getting enough hydrogen sulfide from the fart gases flowing through the air. On the other hand, if you are not around people very much then the only farts you can rely on to preserve your health are your own farts. This means you need to close all your windows, so that the air in your room is trapped where you are. Then every time you fart, you'll have a little bit of hydrogen sulfide in the air that you can breathe. And if you enjoy the smell of your own farts, then more power to you.

9. How to reduce flatulence

All human beings will release gas through their anus because everyone eats and passes food through their digestive system which causes gas to form. But there are ways to reduce the amount of gas that gets generated, so you can reduce flatulence. The most effective way is to eat fewer foods that have high amounts of carbohydrates (or carbs). The three main carbohydrates to avoid are sugar, starch and fiber. Of course, you shouldn't cut these carbohydrates out of your system completely. In fact, it would be impossible to do so because the body depends on carbohydrates for its energy and functionality. You just have to watch how many grams of carbohydrates you are putting into your system. The average adult should not consume more than 200 grams of carbohydrates per day. For kids, from the age of 1 year onwards it is advised by dieticians to eat about 130 grams of total carbohydrates every day.

Anything extra will most certainly cause flatulence and possibly even obesity. However, the carbohydrates you do get should be from healthy foods like fruits, vegetables, wheat and oats. Don't think that buying potato chips or chocolate cookies are going to give you good carbohydrates because they won't. All you will do is ruin your health and fart up a storm.

Most people don't realize that carbohydrates
produce lots of gas in the digestive system. They
even produce more gas than foods that are high in
protein, such as meats. When you have excess
amounts of carbohydrates that are undigested, the
bacteria residing within your gastrointestinal tract
will feed off them. This is what ultimately causes
gas to get produced. You will tend to find
indigestible carbohydrates in foods that have lots of
fiber, like beans and oatmeal. That is why these
foods are known for causing excessive flatulence.
So if you want to avoid farting then cut out these
foods. You should also cut out sweets and breads

because they also have complex carbohydrates that cause flatulence as well. The next food products you need to avoid are animal products, such as dairy and meat. Studies have shown that meat contains hydrogen sulfides, which is a fancy name for sulfur. When people consume meat and it gets broken down in the digestive tract, the sulfur ends up mixing with the gas that eventually comes out of their anus. Once that happens, the farts will end up smelling like rotten eggs because of the sulfur compound in the gas. Therefore, you want to avoid eating meats whenever you can.

Now this book is not meant to be a vegetarian book, but the truth of the matter is that vegetarians have the best smelling farts because they don't eat meat. The only challenge they face is the way they consume their beans and vegetables. After all, if someone is not eating meat then they are going to have to find an alternative source for their protein intake. Beans are one of the best sources of protein for vegetarians that don't want to eat meat. The only other alternative is protein shakes, but those are not recommended because they cause even more flatulence. With that being said, you have to prepare your beans carefully to reduce flatulence. This means washing the beans thoroughly and soaking them for 30 minutes before cooking. The reason you do this is because beans contain indigestible carbohydrates on them called Galacto-oligosaccharides. Soaking the beans will get 25% of these carbohydrates off them because the carbs are water soluble.

When you prepare your vegetarian meal, add some fennel seeds into the food. These seeds are a natural way to fight flatulence in the digestive tract. They go great as a topping for salads and vegetables. Another tip is to chew your food with your mouth closed and try not to talk when you eat. That way you won't end up swallowing air and causing more gas to build up inside you. Also, take small bites when you eat and don't eat too fast. When you are done eating go for a 30 minute walk outside or on the treadmill. This will help your digestive tract move food along faster and reduce the chances of intestinal gas buildup.

But remember, you must eat a balanced diet to stay healthy.

10. Conclusion

Hopefully you haven't gotten too grossed out from reading all of these chapters about flatulence and the different aspects of farting. By now, you should understand that the chemical composition inside our digestive tract is the reason why gas gets created in our intestines and then eventually comes out of us through the anus. It can be kind of weird to think of our bodies as this biological mechanism that produces gases and liquids which break down food and absorbs nutrients. The whole biological system is a true miracle of nature and it should be appreciated for what it is. What's even more amazing is that people think farts are a terrible thing, but they are not. You have already learned in this book that farts are actually healthy to release from your anus. Not only that, they can also be healthy to smell as well, with a little caution of course.

Of course, nature was not perfect when it created
our bodies. We all wish that we didn't have to be
put in embarrassing situations where we
unintentionally fart up a storm and make people
around us uncomfortable by the smell. Either that
or we simply embarrass ourselves because we make
noisy farts that everyone around us can hear. You
might try to pretend the sound came from
something else, like moving furniture or skidding
your shoe on the floor. But you won't be fooling
anyone because they know a fart sound when they
hear one. Remember, people from all walks of life
fart every single day. Nature does not give prejudice
to those who are rich, poor, young or old. All living
human beings fart at least twenty times per day, so
they will recognize the sound of farts when they
hear one. Therefore, just act casual and go about

your business like nothing happen. That is the best advice for what to do in an embarrassing fart situation.

Now you have the knowledge you need about flatulence. You should have a clear understanding of why it occurs in us and how to best manage the smells and sounds of your farts. What you should do is take this newfound knowledge that you have about flatulence and educate people around you about it. After all, you probably have someone in your life that is infamous for farting up a storm.

Perhaps their farts are louder than the farts of anyone else in your life. Instead of just mindlessly thinking of that person as obnoxious or weird, you can now understand the chemical process in their body that is causing these excessive farts to occur. However, you may find it hard to talk to people about flatulence because it is truly a taboo topic in our society. Even though people fart all the time, it is not something that is talked about very much. You may joke about farting when you are a child, but in the adult world it is never discussed. This is probably why few adults understand what flatulence actually is. But now you know about flatulence and that is the important thing. You could recommend this book to those who you feel will benefit from it. The worst that can happen is they feel embarrassed about their flatulence and try to change the subject. Chances are they won't do this though because they are as curious about flatulence as the next person.

Thank you for taking the time to read this book. I would love it if you can leave me feedback on what you have learned about flatulence and how this knowledge will help you manage your own flatulence or the flatulence of those close to you.

Good luck... and may the farts be with you!

11. Further reading and credits

Sophie Le Trionnaire et al. 2014. The synthesis and functional evaluation of a mitochondria-targeted hydrogen sulfide donor, (10-oxo-10-(4-(3-thioxo-3H-1,2-dithiol-5-yl)phenoxy)decyl)triphenylphosphonium bromide (AP39). Med. Chem. Commun: 5, 728-736

Wynn Kapit and Lawrence M. Elson. 2013. The Anatomy Coloring Book.

Patricia J. Wynne, Donald M. Silver. 2009. My First Human Body Book (Dover Children's Science Books).

Tish Rabe, Aristides Ruiz. 2003. Inside Your Outside: All About the Human Body (Cat in the Hat's Learning Library).

Patty Carratello. 2004. My Body (Science Books).

Nishi Singh. 2014. Cells For Kids (Science Book For Children).

Carbohydrates. National Institutes of Health.

Crai S. Bower, Travis Millard. 2008. Farts: <u>A Spotter's Guide</u>.

Reepah Gud Wan. 2003. <u>The Zen of Farting</u>.

Jim Dawson. 1998. <u>Who Cut the Cheese?: A Cultural History of the Fart</u>.

Alec Bromcie. 2011. <u>The Complete Book of Farts</u>.

***

Made in the USA
Middletown, DE
30 May 2019